U.S. CHEMICAL SAFETY AND HAZARD INVESTIGATION BOARD

INVESTIGATIVE STUDY

Final Report

PUBLIC SAFETY AT OIL AND GAS STORAGE FACILITIES

I0502966

MULTIPLE SITES
(44 FATALITIES, 25 INJURED)
26 INCIDENTS FROM
1983-2010

KEY ISSUES:

- OIL AND GAS EXPLORATION AND PRODUCTION FACILITIES PRESENT HAZARDS TO MEMBERS OF THE PUBLIC INCLUDING CHILDREN

- SECURITY MEASURES ARE INSUFFICIENT AT EXPLORATION AND PRODUCTION FACILITIES

- REGULATIONS AND INDUSTRY STANDARDS DO NOT PROVIDE UNIFORM, EFFECTIVE GUIDANCE

REPORT NO. 2011-H-1
SEPTEMBER 2011

Table of Contents

Abbreviations

API	American Petroleum Institute
bbl	Barrel (42 U.S. gallons)
CSB	U.S. Chemical Safety Board
E&P	Exploration and Production
EIA	U.S. Energy Information Administration
EPA	U.S. Environmental Protection Agency
GOR	Gas Oil Ratio
MSO&GB	Mississippi Oil and Gas Board
NFPA	National Fire Protection Association
OCC	Oklahoma Corporation Commission
OSHA	U.S. Occupational Safety and Health Administration
psig	Pounds per square inch gauge
RRC	Railroad Commission of Texas

1.0   Executive Summary

1.1   Oil and Gas Storage Sites Present a Hazard in Rural Areas

On October 31, 2009, two teenagers, aged 16 and 18, were killed when a petroleum storage tank exploded in a rural oil field in Carnes, Mississippi. Six months later a group of youths were exploring a similar tank site in Weleetka, Oklahoma, when an explosion and fire fatally injured one individual. Two weeks later, a 25-year-old man and a 24-year-old woman were on top of an oil tank in rural New London, Texas, when the tank exploded, killing the woman and seriously injuring the man. In April 2010, the U.S. Chemical Safety Board (CSB) initiated an investigation into the root causes of these tragic incidents. All three incidents involved rural unmanned oil and gas storage sites that lacked fencing and signs warning of the hazards, which might have otherwise deterred members of the public from using them as places to gather.

Oil and gas storage sites are part of the landscape in many rural American communities and an important component of the country's vast system of oil and gas exploration and production. Over 800,000 crude oil and natural gas producing facilities are distributed across the U.S., often located in wooded clearings or other isolated locations.

However, in many states, these sites can be placed as close as 150 to 300 feet from existing residences, schools, churches and other structures. Only in a few large cities where these sites exist – Houston, Oklahoma City, and Los Angeles – are constraints placed on the location of the facilities within the city limits.[1]

In most cases, however, these sites are away from public view, often unfenced, unsupervised, and lacking warning signs. They have proven to be a tempting venue for young people looking for a place to gather,

---

[1] New Jersey Petroleum Council and The American Petroleum Institute. Oil and Natural Gas Industry Security Assessment and Guidance. 2002.
    <http://www.nj.gov/dep/rpp/brp/security/downloads/NJ%20Best%20Practices%20Petroleum%20Sector.pdf>

and socialize.   Activities where an ignition source is introduced into the tank, or even the presence of static electricity or lightning, can cause hydrocarbon vapors in the tanks to ignite and explode.

1.2   CSB Study

To prevent future deaths and injuries, the CSB investigated the root causes of the three incidents and conducted an analysis of the regulatory framework that contributed to the prevalence of this type of event. The CSB examined federal, state, and local regulations; inherently safer designs of tanks; and industry standards and practices recommended by the American Petroleum Institute (API) and the National Fire Protection Association (NFPA).   The CSB also administered a survey to gauge the public's view of these sites and the issues arising from their presence in the community.   Among 190 survey recipients in a rural Mississippi community, 11 percent of respondents stated they had "hung out at oil sites."   When asked about the type of activity engaged in at oil sites, 14 percent stated they socialized; 19 percent said that they rode four-wheelers at oil sites; and 11.5 percent stated that they hunted.

This CSB study provides recommendations to strengthen security at exploration and production (E&P) sites to include fencing, warning signage, locking of all hatches and using inherently safer tank features to prevent future incidents.

1.3   Findings from Oil and Gas Site Incidents

The CSB found the three explosions in Mississippi, Oklahoma, and Texas could have been prevented or made less likely by restricting access to the facilities, by providing warning signage, by securing the hatches on the tanks or utilizing inherently safer tank design at these facilities. The growing number of oil and gas facilities nationally, their accessibility to members of the public, and the lack of awareness among the public about the hazards posed by the tanks suggest a potential for similar incidents.   The CSB makes the following key findings:

1. Members of the public, most often children and young adults, commonly visit oil and gas production sites without authorization for recreational purposes.

2. Members of the public gain access to production tanks via attached unsecured ladders and catwalks, and may come into contact with flammable vapors from tank vents or unsecured tank hatches.

3. Members of the public, unaware of the explosion and fire hazards associated with the tanks, unintentionally introduce ignition sources for the flammable vapor, leading to explosions.

4. The CSB identified 26 similar incidents between 1983 and 2010, which resulted in a total of 44 fatalities and 25 injuries. All the victims were 25 years of age or less.

5. The three incidents investigated by the CSB in 2009-2010 occurred in isolated, rural wooded areas at production sites that were unfenced, did not have clear or legible warning signs, as required under OSHA's Hazard Communication Standard, and did not have hatch locks to prevent access to the flammable hydrocarbons inside the tanks.

6. The storage tanks did not include inherently safer design features to prevent tank explosions. Safer design features used in the downstream, refining sector would likely prevent tank explosions at E&P sites. These include the use of vents fitted with pressure-vacuum devices, flame arrestors, vapor recovery systems, floating roofs or an equivalent alternative.

7. E&P storage tanks are exempt from the security requirements of the Clean Water Act and from the risk management requirements of the Clean Air Act.

8. Industry guidance from the American Petroleum Institute recommends specific security measures for storage tanks of refined petroleum products but not for storage tanks at upstream E&P sites, and the National Fire Protection Association standards do not adequately define security expectations where these deadly incidents occurred.

9. Some states, including California and Ohio, and some localities have mandated security (including fencing, locked or sealed tank hatches, and warning signs) for E&P sites, particularly in urban areas. As a result, despite its large role as an oil producing state with many of these types of facilities, none of the 26 incidents occurred in California. However, many other large oil and gas producing states have no such requirements (except for certain E&P sites where toxic hydrogen sulfide gas is present).

1.4   Recommendations

As a result of the findings from this study, the CSB makes recommendations to the following recipients:

- U.S. Environmental Protection Agency (EPA)

- Mississippi Oil and Gas Board

- Oklahoma Compact Commission

- Texas Railroad Commission

- American Petroleum Institute (API)

- National Fire Protection Association (NFPA)

2.0   Hazard at Oil and Gas Production Facilities

Between October 2009 and April 2010, four teenagers and young adults lost their lives from explosions at three different oil and gas production sites in rural Mississippi, Oklahoma, and Texas. The CSB first became aware of this hazard in 2003 when a similar explosion in Palestine, Texas, fatally injured four teenagers. In 2010, the CSB initiated an investigation to further examine the issue. The CSB found 26 similar incidents involving explosions and fires at oil and gas production sites.

2.1   CSB Outreach

After the CSB's initial deployment to a tank explosion in Carnes, Mississippi, the agency created a safety video targeting individuals under the age of 25 to increase awareness of the hazards posed by oil and gas sites. The safety video, "No Place to Hang Out: The Danger of Oil Sites," incorporates the experiences of the victims' friends, families, and community leaders in Carnes, and is intended to be integrated into high school and middle school curricula. The CSB distributed this video to school superintendents across Mississippi and continues to work with safety advocates in an effort to reach young people who live in oil and gas producing communities.

2.2   Study Methodology

To further understand why these incidents were occurring across the country, the CSB deployed to and investigated the three oil and gas tank explosions discussed above and collected information on 23 similar explosions across the country. Investigators interviewed key witnesses and first responders at each of the three incident sites and gathered exploration and production (E&P) site records from each state oil and gas regulator. Incident reports for the 23 additional incidents were requested from local responders. A survey of high school students and community members in Carnes, Mississippi, was administered to understand the use of and hazard awareness at oil and gas facilities. The CSB then analyzed safety and security regulations at the local, state, and federal level as well as relevant industry standards in order to identify systemic gaps and formulate recommendations aimed at preventing future incidents.

3.0   Characteristics of Oil and Gas Storage Facilities

3.1   Process Overview

At typical E&P sites, crude oil and natural gas are pumped from underground hydrocarbon reservoirs to the surface.   The well stream is connected to a piping system that transports hydrocarbons to an oil-gas separator where gas and water are removed from crude oil.   The oil is then transferred to storage tanks in a tank battery[2] until it is pumped into a transport truck for eventual sale (Figure 3-1).

In states where vapor recovery systems[3] are not mandated,[4] oil tanks are usually equipped with a tank hatch[5] and an atmospheric vent on the surface (Figure 3-1). Oil field workers regularly check liquid levels through the hatch, which is accessible by a walkway or catwalk.[6] The oil-gas separator also contains an atmospheric vent that releases hydrocarbon vapors. 210-barrel capacity atmospheric storage tanks – which were involved in two of the three explosions investigated by the CSB – are commonly used to store crude oil and condensate at E&P facilities throughout the U.S. These tanks are rated for petroleum liquids with a vapor pressure of less than 0.5 psig[7] and are selected "based on vapor pressure, flash point, potential for explosion, temperature and specific gravity."[8] If circumstances change inside the tank and

---

[2] A tank battery is an installation of several tanks at E&P facilities.

[3] A vapor recovery system consists of a sealed vapor gathering system capable of collecting the hydrocarbon vapors and gases discharged and a vapor disposal system capable of processing such hydrocarbon vapors and gases so as to prevent their emission into the atmosphere.

[4] California state law requires oil and gas sites located in non-attainment (non-compliant) air pollution areas to capture all hydrocarbon vapors produced in a stock tank and cycle them through a vapor recovery system.

[5] A tank hatch is a covered opening on the surface of a tank.

[6] A catwalk is the stair or ladder leading to and providing access to the top of a tank or vessel.

[7] Myers, P.E. Aboveground Storage Tanks. 1997. NY: McGraw-Hill, p. 25.

[8] *Ibid.*

the internal pressure increases significantly above its pressure rating, the tank loses its structural integrity and fails.

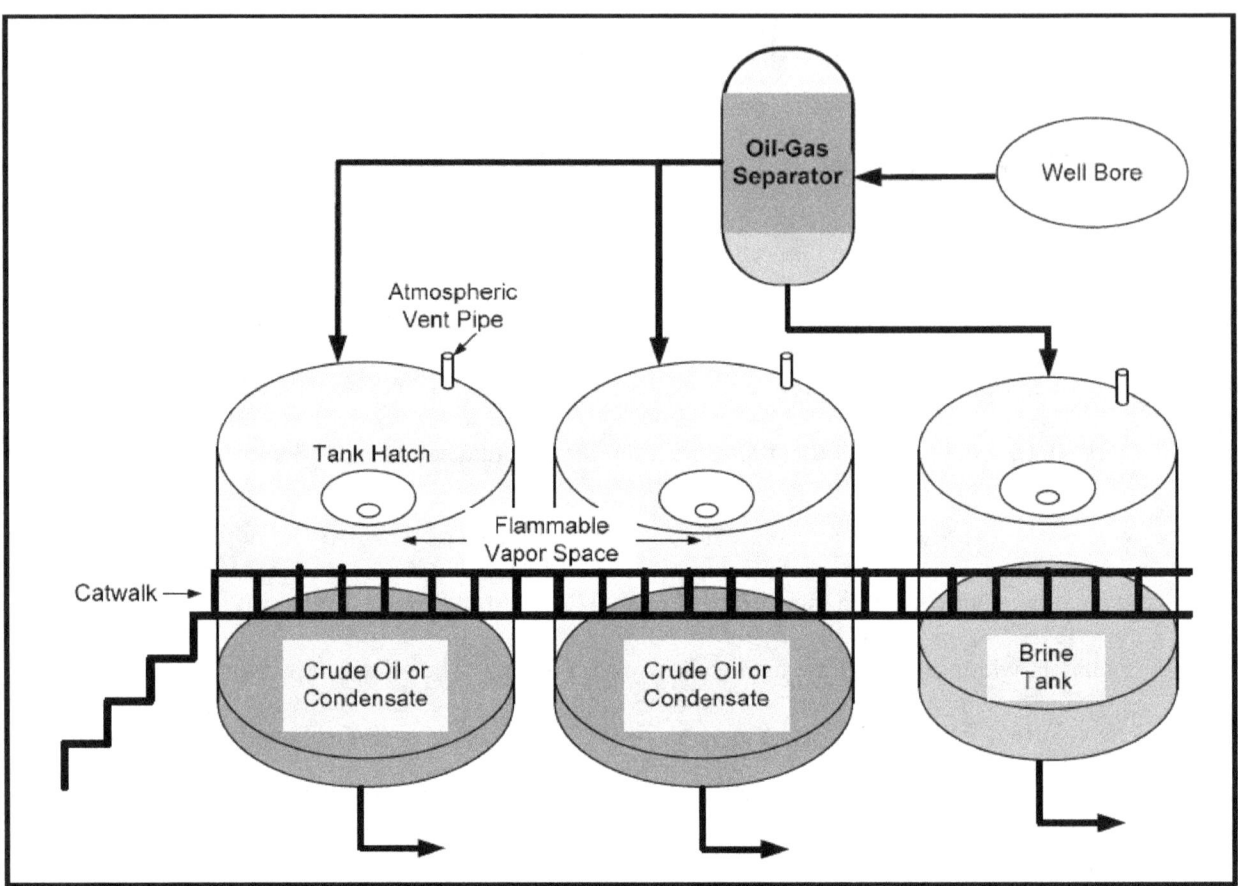

Figure 3-1: Basic schematic of oil and gas production facility.

3.1.1   Measuring Flammability Properties at Oil and Gas Facilities

The flammability of the product at E&P facilities varies depending on the geology of the formation and can change over time. The flammability of crude oil and condensate may be characterized by specific gravity (weight per volume), American Petroleum Institute (API) gravity, and the gas oil ratio (GOR). The higher the API gravity, the lighter and more flammable the compound; materials below an API gravity value of 35 are characterized as crude oil, while those above 45 are considered condensate. Light crudes generally exceed 38 degrees API and heavy crudes have an API gravity of 22 degrees or below.

Intermediate crudes fall in the range of 22 degrees to 38 degrees API gravity. API gravity is measured using stock oil taken from the storage tanks and is often reported to state oil and gas boards upon initial production from a well. The Gas Oil Ratio (GOR) measures the dissolved natural gas remaining in a well stream at a specific pressure and temperature and is a ratio of the gas produced for each barrel of stock oil in standard cubic feet per barrel (scf/bbl).

In addition to being flammable, crude oil and associated produced water (also referred to as brine) may contain varying levels of hydrogen sulfide[9] depending on the geology of the hydrocarbon reservoir. Crude oil storage sites containing hydrogen sulfide are typically subjected to stronger regulatory requirements.

3.2   Close Proximity of Oil and Gas Facilities to Residents

In most states oil and gas leases are divided into a mineral[10] and surface estate[11]. In the past, both the surface and mineral estates were transferred when a property was sold. It is now common for mineral estates and surface estates to be severed and sold separately. However, federal and state laws allow dominance of mineral estate rights over the rights of the surface estate. This supremacy allows mineral estate owners to lease their rights to oil and gas operators, who utilize the surface estate to access the minerals beneath the surface.[12] Although requirements vary across states, surface estate owners can refuse access to the minerals estate owners or charge a fee for access. While some states may institute "accommodation" statutes that enable a surface estate owner to reduce the impact of the exploration activities to the surface, such an agreement is not required by an operator who has leased the mineral

---

[9] Hydrogen sulfide is hazardous and deadly at low concentrations with an Immediately Dangerous to Life and Health concentration of 100 parts per million-(ppm) in air.

[10] Mineral estate refers to the ownership of mineral rights or "mineral interest"-- all unusual organic and inorganic substances in the soil giving it value on a property.

[11] Surface estate refers to the ownership of the surface of a property above the mineral estate.

[12] Texas Rail Road Commission. Oil & Gas Exploration and Surface Ownership.
< http://www.rrc.state.tx.us/about/faqs/SurfaceOwnerInfo.pdf>

estate from its owner.[13] Additionally, once an oil or gas operator obtains a mineral lease, drilling operations can occur in close proximity to existing residences, without notifying the surface owner.

In both urban and rural areas, drilling may occur within a few hundred feet of residences. For example, in many states, minimum requirements for the placement of oil sites and tank storage facilities range from 150 to 300 feet from existing residences, schools, churches, and other structures.[14] CSB investigators observed that a number of oil and gas production facilities in Mississippi, Oklahoma, and Texas are unsecured and located within rural communities and in close proximity to residences. The CSB also learned that although the mineral estate can be leased to an oil or gas operator, unless stipulated otherwise in the leasing agreement, the surface estate owner can concurrently lease the surface rights of the same property as part of hunting leases or for other uses.

---

[13] Court of Appeals of Mississippi. No. 2003-CA-01572-COA. *MS: Turner vs EOG Resources.*
<http://caselaw.findlaw.com/ms-court-of-appeals/1243004.html>

[14] Colorado Law Institute. Comparison of State Oil and Gas Regulations and Local Ordinances Regarding Setbacks for the Intermountain West. 2009.
<http://www.oilandgasbmps.org/laws/setback.standards.comparison.10.8.09.pdf>

4.0   Increase in Oil and Gas Storage Sites Poses Increased Risk

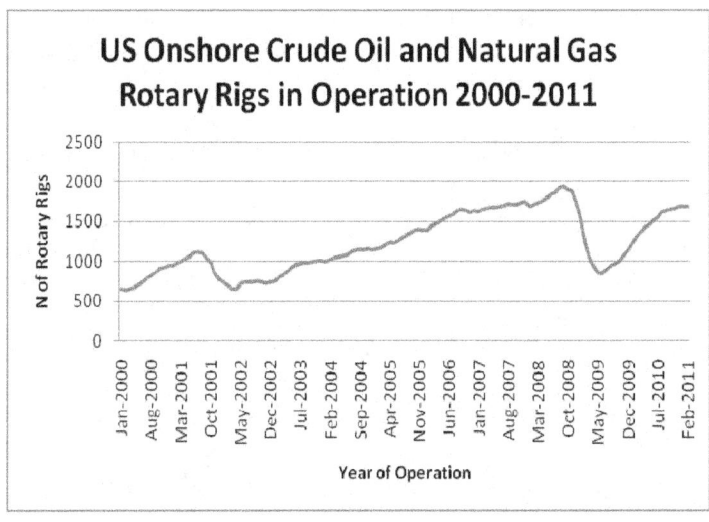

4-1: EIA U.S. Onshore Crude Oil and Natural Gas Rotary Rigs in Operation

As the number of oil production sites and the population density increase, so does the likelihood that young people will access oil sites as places to "hang out." In 2009, the Energy Information Administration (EIA) reported that there were at least 363,459 active oil and 461,388 active gas well sites throughout the U.S.[15] Approximately 85 percent of oil and gas wells are small producers generating 15 barrels of oil equivalent per day (BOE/day)[16] or less.[17]

The EIA data demonstrate a general increase in drilling activity over the past decade (Figure 4-1). In addition, drilling in shale for natural gas exploration and development has nearly doubled from 2009 to 2010 and active wells increased from 11,657 to 20,388.[18] The increase in oil and gas drilling activity for crude oil and natural gas exploration creates an increase in the number of oil and gas production sites, likely increasing the risk to members of the public.

---

[15] The Energy Information Administration. United States Total 2009, Distribution of Wells by Production Rate Bracket. < http://www.eia.gov/pub/oil_gas/petrosystem/us_table html>

[16] Barrels of Oil Equivalent per day is used in the production or distribution of oil. One barrel of oil has the same amount of energy content as 6,000 cubic feet of natural gas.

[17] The Energy Information Administration. United States Total 2009, Distribution of Wells by Production Rate Bracket. <http://www.eia.gov/pub/oil_gas/petrosystem/us_table html>

[18] The Energy Information Administration. Annual Energy Outlook 2011.

<http://www.eia.gov/forecasts/aeo/pdf/0383(2011).pdf>

5.0   Oil and Gas Tank Explosions in the U.S. from 1983 to 2010

Through media searches, the CSB identified 26 similar incidents that occurred from 1983 to 2010 at oil

and gas production sites in 10 different states (Figure 5-1). These incidents resulted in 44 fatalities and 25

injuries to members of the public under 25 years of age (Appendix A). The majority of these incidents

occurred in rural areas, where the sites lacked security and safety measures such as fencing, warning

signs, or locks on tank hatches. The CSB collected investigation reports from oil and gas boards, local

fire departments, and/or state environmental agencies detailing the circumstances and consequences of

these incidents (Appendix A). The reports illustrate the explosion hazard to members of the public who

wander into these sites for recreational purposes. The data are limited to accidents covered in the media,

since a central database for tracking incidents involving members of the public does not exist. For this

reason, a background rate of the frequency of these incidents could not be obtained. However, the CSB

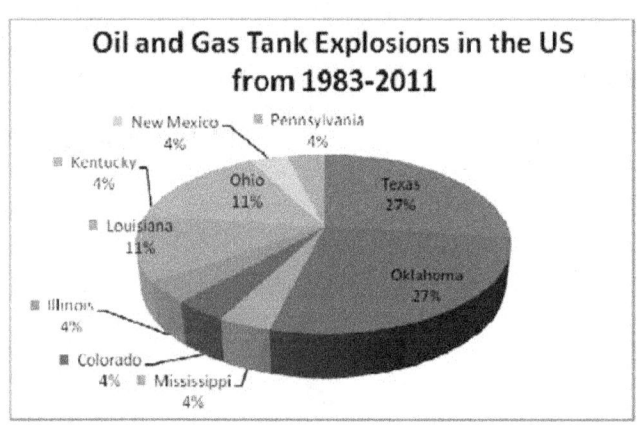

Figure 5-1: Oil and Gas Storage Tank Explosions
across 10 states, 1983-2011

found these incidents are occurring consistently, although the data were not sufficient to demonstrate a meaningful trend.

Of the incidents reviewed, the CSB concluded that 82 percent of the fatally injured victims were teenagers and 18 percent were young adults between the ages of 20-25. Sixty-nine percent of the incidents involved multiple injuries or fatalities.

The majority of the 26 incidents occurred in Texas (27 percent) and Oklahoma (27 percent); however the

remaining 46 percent of incidents occurred in oil and gas production states throughout the country (Figure

5-2). The CSB discovered approximately 84 percent of the 26 incidents occurred in areas that did not

have any state or local zoning ordinances that required security fencing, signs, or hatch locks to discourage site access. Only 16 percent of the incidents occurred in areas where zoning ordinances appeared to require fencing for sites in urban locations. (See Appendix A for more details on incidents).

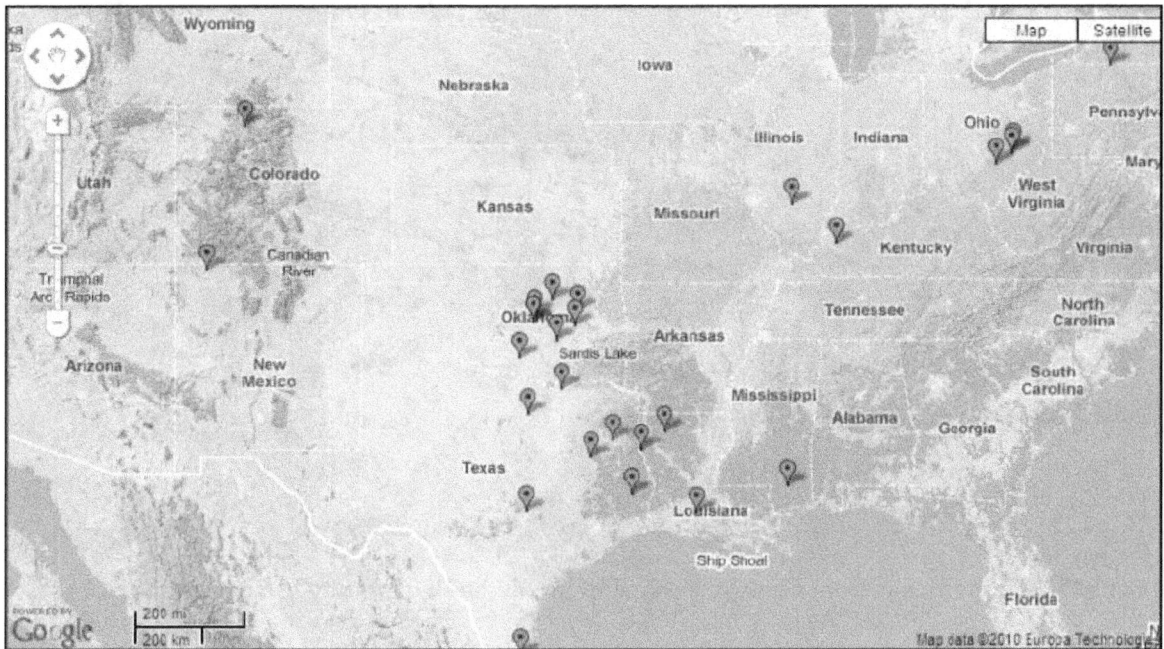

Figure 5-2: Map of oil and gas production facility explosions from 1983-2010 that killed or injured members of the public.

6.0   CSB Investigations, 2009-2010

The CSB investigated the three recent incidents described in this section to develop a more thorough understanding of why tank explosions continue to occur at E&P sites.

17

6.1   Delphi Oil, Carnes, Mississippi

Figure 6-1: Delphi Oil gas production site, Carnes, Mississippi

6.1.1   Incident Details

On October 31, 2009, two teens from Carnes, Mississippi, were fatally injured when a tank of gas distillate exploded at a rural oil and gas storage site located near several residences. The two teenagers, 18 and 16, arrived at the home of one of the victims at approximately 11:00 p.m. on October 30, 2009. Between midnight and 4 a.m., the victims drove to the adjoining gas well site located approximately 150 yards from the home in a nearby clearing in the woods. At approximately 4:00 a.m. a violent explosion occurred inside one of the site's two storage tanks.

Figure 6-2: A) Remnants of exploding tank
B) Tank hatch located approximately 300ft from berm
C) Bottom of tank found 60 feet from berm in adjacent wooded area

The force of the explosion propelled the upper part of the tank approximately 225 feet while the bottom of the tank was thrown about 60 feet in the opposite direction. The tank's vent pipe and hatch detached from

Figure 6-3: Delphi Oil gas production site identification sign

the tank top and landed over 300 feet away (Figure 6-2B). The exploding tank lost all its contents, triggering a large fire up to 200 feet high that persisted for about four hours. The fire prevented first responders from accessing the sign containing Delphi Oil's emergency contact information, which was located at the well head within the fire zone (Figure 6-3). Both teenagers were killed instantly; their bodies were found approximately 120 feet from the original location of the tank. Forrest County Sheriff photographs taken immediately following the incident demonstrate the two victims were thrown from the catwalk. Although a cigarette lighter was found at the site there was no evidence that it was the ignition source.[19] CSB staff traveled to

---

[19] Ignition sources may include matches, lighters, cigarettes, lightning, static electricity, and in some cases pyrophoric iron sulfide.

the incident site in November 2009 and January 2010 and interviewed emergency responders, neighbors, family members, and friends of the victims.

6.1.2   Incident Site

Figure 6-4: Delphi Oil gas production site surrounded by woods

Figure 6-5: Oil and gas production site in Carnes, Mississippi with elevated catwalk

The explosion and fire occurred at an active natural gas well site leased by Delphi Oil, an oil and gas producer based in Baton Rouge, Louisiana. Delphi Oil began exploring and developing the site in November 2006, when it obtained the mineral leases to a number of lots in Forrest County. Delphi Oil leased approximately 400 acres of mineral rights for the Delphi Oil 5-18 No. 1 gas production well. The site produced natural gas and gas condensate,[20] a mixture of light hydrocarbons, which was stored in a 210-barrel[21] capacity tank interconnected to an adjacent 210-barrel tank that stored brine. Each tank had a six-inch diameter hatch and a vent pipe located on the roof that was open to the atmosphere.

6.1.3   Flammability

At the time of the incident, the exploding tank contained approximately 14 barrels of condensate. According to records from the Mississippi Oil and Gas Board (MSO&GB), the distillate from the well had an API gravity of 50 degrees, indicating the presence of a highly volatile hydrocarbon mixture.

---

[20] Condensate: A flammable natural gas liquid recovered from gas wells using separation equipment.

[21] A barrel is a unit of volume equal to 42 U.S. gallons.

6.1.4   Access, Fencing, Warning Signs, and Security

This particular gas well site is located in a rural clearing surrounded by woods (Figures 6-1 and 6-4) approximately 500 feet from an adjacent residential property. It is readily accessible from a number of foot trails in the woods and an unsecured dirt access road. At the time of the explosion, the site did not have signage to warn of the hazardous contents of the tanks, hatch locks, perimeter or equipment fencing to deter public access, nor were the tanks designed to avoid an internal vapor explosion.

6.1.5   Recognition of Oil and Gas Site Hazards

Rural oil and gas tank production sites are often in remote locations that are cleared for the installation of extraction and storage equipment. Residents and friends of the victims told CSB investigators that prior to the incident, they did not view oil and gas production sites as dangerous, as they were part of the landscape. Many had grown up in close proximity to oil rigs and tank storage sites and used them as common gathering locations for recreational purposes such as socializing, hunting, and driving all-terrain vehicles. The CSB learned that a number of hunting leases in the area incorporate oil and gas sites and their storage facilities, and some residents described using the sites' elevated catwalks (Figure 6-5) and the tanks as platforms for hunting.

6.2   Three MG Family, Weleetka, Oklahoma

Figure 6-6: Weleetka, OK, oil and gas well site indicating Tank 4-22, which exploded

6.2.1   Incident Details

On April 14, 2010, a 210-barrel capacity tank exploded at an oil and gas production site in rural

Weleetka, Oklahoma, fatally injuring a 21-year-old male and causing second-degree burns to a 26-year-

old male. At the time of the explosion, a group of six young adults and teenagers, ages 18 to 32, were

socializing at the oil and gas site. They were on their way to an isolated location along the North

Canadian River in Oklahoma, when they turned off onto a rural dirt road, which also provided access to

the oil and gas site. According to witness testimony, rather than continue directly to the riverbank, the

group stopped at the oil and gas site about a half mile from their intended destination at approximately

8:30 p.m. Shortly thereafter, witnesses said the victim ascended stairs to the catwalk that accessed the

three oil storage tanks belonging to Three MG Family, Inc. Witnesses said that the victim looked into the

hatch of tank 4-22, possibly while smoking; at the same time another friend was walking behind him in

the dark and struck his lighter to see. Vapors from the tank ignited and an explosion ensued.

A fireball engulfed the victim, causing third degree burns covering up to 95 percent of his body. The

Weleetka Fire Department received a 9-1-1 call at 9:06 p.m. and, upon arrival at the scene, observed a

Figure 6-7: A) Gate at entrance to dirt road leading to Weleetka oil and gas well site B) Cattle gate on dirt road leading to oil and gas well. C) Sign identifying gas pipeline at Weleetka oil and gas site

raging fire. The victim was airlifted to a Tulsa burn center but succumbed to his injuries the next day. The exploding tank (Tank 4-22) contained approximately 155 barrels of crude oil; the adjacent tank had approximately 10 barrels of crude oil and the tank closest to the oil separator was half-full of brine (Figure 6-6). Three Enterprise Oil tanks on the opposite side of the oil-gas separators were unaffected by the explosion and ensuing fire. As a result of the explosion and fire, oil spilled around the tanks and fire spread about 15 feet onto the surrounding brush area behind the affected tanks. The site did not have a berm or dike to contain the oil released from the exploding tank when it lost its contents.

6.2.2   Incident Site

The oil and gas site was leased by Three MG Family, ScissorTail Energy, LLC, and Enterprise Energy. The site contained six 210-barrel capacity storage tanks, two oil separators, and two gas separators. One set of oil and gas separators was connected to three adjacent interconnected storage tanks belonging to Three MG Family, Inc.; the other set provided oil and gas to three interconnected storage tanks belonging to Enterprise Oil (Figure 6-6).

6.2.3   Flammability

Three MG Family, Inc. sales records show that the three interconnected tanks were emptied on April 10, 2010, four days before the explosion and fire, leaving a significant vapor space above the remaining contents in the tank. According to company transport documents, the crude oil in the exploding tanks had an API gravity rating of approximately 39.0, falling between the range of crude oil and condensate. A transport receipt dated March 4, 2010, identified the contents of the affected tanks as Petroleum Crude Oil, 3 UN 1267. A material safety data sheet (MSDS) for this type of oil describes the material as "easily ignited by heat, sparks or flames." Section 4 of the MSDS, "Fire and Explosion Hazards," states that "vapor/gas will spread along the ground [and] collect in low or confined areas (sewers, basements, tanks). [Vapors] may also travel to a source of ignition and flash back. Containers may explode when heated."[22]

6.2.4   Access, Fencing, Warning Signs, and Security

The oil tank site involved in this incident is located in a wooded clearing less than a mile from the main road and half a mile from the banks of the Northern Canadian River. The dirt road leading to the site is unlit and secured by a typically unlocked iron cattle gate located where the dirt road intersects the main road (Figure 6-7A, B). The gate is approximately 4 feet high and 12 feet long and is the only means of discouraging access to the site (Figure 6-7B). The site lacks a perimeter fence, warning signs identifying hazards of the flammable materials inside the tanks, or hatch locks. The design of the failed tank did not prevent an internal explosion. There was one warning sign identifying the location of a gas pipeline on site (Figure 6-7C). The site is unmanned except for a well tender who checks the oil levels in the tanks

---

[22] Irving Oil. MSDS Crude Oil.
<http://www.irvingoil.com/dloads/refinery/03050%20CRUDE%20OIL%20MSDS.pdf.>

each morning. The CSB determined from witness testimony that the gate to the dirt road leading to the oil

tank storage site was often left unlocked and on the day of the incident was likely unlocked.

6.3   MC Production, New London, Texas

Figure 6-8: MC Production Oil tank site, New London, Texas

6.3.1   Incident Details

At approximately 1:00 a.m. on April 26, 2010, an oil tank exploded in New London, Texas, fatally

injuring a 24-year-old woman and seriously injuring a 25-year-old man. The exploding tank was

propelled 48 feet away from its original location; the top of the tank was found 35 feet from its original

location.   The CSB learned that two individuals were climbing the stairway of the catwalk when one

victim asked the other to light a cigarette. Witness testimony revealed that when the second victim lit the

cigarette, an explosion ensued.

6.3.2   Incident Site and Flammability

The oil and gas site was leased by MC Production. At the time of the explosion, there were three

interconnected 1000-barrel capacity tanks at the facility. The exploding tank contained a small amount of

hydrocarbons and another adjacent tank had a hole. The oil site has been in operation for at least 80

years. According to well records obtained from the Texas Railroad Commission (RRC), the oil lease was

active at the time of the explosion producing 185 barrels of oil and selling 369 barrels during the month of

the incident. However, testimony from the well tender indicates the oil tank involved in the explosion

Figure 6-9: A) Open cattle gate leading to storage site B) Access road leading to tank storage site
C) Woods surrounding oil tank site

had not stored product for at least one-and–a-half years prior to the explosion. The CSB also learned that
MC Production reported the oil from the site had an API gravity of 37.4, characteristic of an intermediate
crude oil that can produce flammable hydrocarbons.

6.3.3   Access, Fencing, Warning Signs, and Security

Figure 6-10: M-C Production tank site with
warning sign

The MC Production oil field is located at the end of an isolated
road in the middle of a clearing surrounded by woods (Figure 6-
9C). According to Rusk County Fire Department officials and
the Rusk County Sheriff's office, at the time of the explosion
the oil site had no fences or hatch locks nor were the tanks
designed to reduce the potential of an internal explosion. A
cattle gate marked the entrance to a dirt road that led to the tank
battery site over 200 feet away (Figure 6-9A). The site did
have one warning sign covered by graffiti; however, its exact
location at the time of the incident is unclear. Witness

testimony revealed the sign may have been moved three to four times the day after the explosion.

Although the "No" on the sign was blurred from graffiti, it warned against smoking, matches or open

lights (Figure 6-10).

Incident photos also revealed the tank site contained graffiti tags from local gangs (Figure 6-10). In addition, on the night of the explosion, a pink children's bicycle was found at the tank site (Figure 6-11). Both the presence of the graffiti and the children's bicycle indicate the tank site was visited by various members of the public. Two days after the fatal accident, Rusk County investigators returned to the accident scene to find a new steel gate with locks and signs at the entry to the access road.

Figure 6-11: A) Exploding Tank at MC Production oil site B) Children's bicycle found on M-C Production tank site on the day of the explosion.

6.4   Recreational Use of Oil and Gas Sites

6.4.1   Survey Methods

The three tank storage sites investigated by the CSB were located in rural areas in close proximity to residential communities. To further assess public understanding of oil and gas site hazards, the CSB conducted a survey of students at Forrest County Agricultural High School, where the victims of the October 2009 explosion were enrolled, and other members of the Carnes, Mississippi, community. The survey was conducted during the spring of 2010, several months after the explosion. The surveys were administered by Forrest County Agricultural High School personnel. Participants were asked to provide age and gender information but no other personal identifiers. A total of 190 surveys were completed; participants had a median age of 16. The survey results are summarized below.

6.4.2   Survey Results

Similar to CSB interviews of community members, the survey results reveal that many local residents (especially children and young adults) view oil and gas sites as convenient places to gather and participate in recreational activities, made easier by relatively unhindered access. Respondents to the survey stated overwhelmingly that they would avoid these sites if hazard signs were present or if access were made more difficult with perimeter fencing and locks.

In the survey, 11 percent of respondents stated they previously "hung out at oil sites." When asked about the activity engaged in at an oil or gas production site, 14 percent stated they spent time with friends; 19 percent said that they rode four-wheelers; and 11.5 percent stated that they fished or hunted. Seven respondents said they had climbed onto the catwalk at an oil site, while six stated they consumed alcoholic beverages and four stated that they smoked cigarettes or cigars at oil sites.

Of the 190 respondents, 11 indicated they visited oil sites once a year and 21 said they did so less than once a year. Five respondents stated that they had lifted the hatch of an oil storage tank; and seven stated they used a lighter or a match while at an oil well site.

7.0   Inherently Safer Tank Design

Inherently safer tank design could have prevented the formation of a flammable atmosphere inside the tanks and likely prevented the three incidents investigated by the CSB, as well as the 23 other similar tank explosions that were identified. The following are examples of tank design features that can be used at E&P facilities to isolate and contain the flammable vapors in order to prevent a vapor space explosion.

An internal floating roof is a design feature where a roof floats on top of a flammable liquid, reducing the hydrocarbon vapor to low concentrations well below the flammable limit. In the past, smaller diameter tanks (e.g. less than 30 feet diameter) could not practically use floating roofs because of stability issues. Today, due to API 650 relaxed buoyancy requirements for small tanks and the development of new composite floating roof materials, floating roofs can be installed in new or existing tanks as small as 8-10 feet in diameter. Currently, most E&P storage tanks have fixed roofs—a less costly alternative to an internal floating roof.

A second inherently safer design feature is the use of pressure vacuum (PV) relief valves. Pressure vacuum relief valves are commonly used on fixed roof tanks to minimize evaporation losses. However, they effectively isolate ignition sources, essentially acting as flame arrestors, so that external ignition sources nearby will not flash back to the vapor space, causing a tank explosion. The valves are designed to prevent the accumulation of pressure or vacuum which could compromise the tank integrity. However, most existing E&P oil storage tanks use open vents when storing flammable liquids. Only those tanks located in areas with strict air pollution rules (e.g. in California) avoid the use of open atmospheric vents. The likelihood of a flash back can be significantly reduced by the use of PV relief valves.

A third design option (one which is recommended for tanks located in urban areas of Ohio) is the use of an actual flame arrestor—a device that extinguishes a developing flame outside a tank, preventing it from entering the vapor space. The flame arrestor forces a flame front through narrow channels that inhibit the propagation of the flame. Both flame arrestors and pressure vacuum valves are similar in function in that they act as barriers to flame propagation from outside the tank into the vapor space.

A final option is the use of a vapor recovery system—a closed system that keeps flammable vapors inside the tank. This system requires the entire tank (tank hatches, atmospheric vents and all tank orifices) to be sealed and isolated from the atmosphere, thus preventing external ignition sources from entering the vapor space. The internal vapors are either recovered for future use or routed to a flare system. Vapor recovery systems are required for tanks located in poor air quality zones in California.

8.0   Oversight of Security at Oil and Gas Storage Facilities

The CSB reviewed federal, state, and local regulations as well as industry standards and guidance to evaluate the existing safety and security requirements for preventing public access to oil and gas production sites.

8.1   Mississippi Oil and Gas Rules

8.1.1   County Rules: Oil and Gas Well Sites in Forrest County, Mississippi

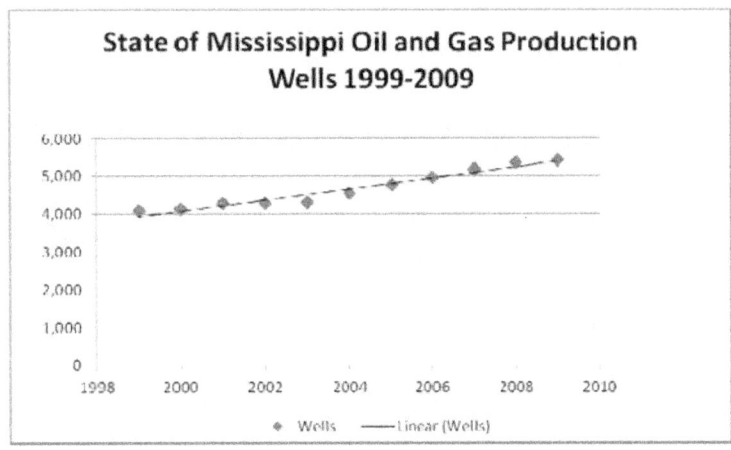

Figure 8-1: Oil and gas production wells in Mississippi (Source: Mississippi Oil and Gas Board)

The Delphi Oil gas site involved in the October 31, 2009, explosion is located in the Pistol Ridge oil field. Delphi Oil leases 15 of the producing wells and their associated tank storage sites in the area. In 2010, following the incident, the Forrest County Emergency Management District (FCEMD) conducted an analysis and reported 119 oil and gas production and storage sites in Forrest County, which includes the rural community of Carnes. Of the 119 oil and gas sites, only 15 were fenced at the time of the FCEMD report.[23] At the time of the incident, Forrest County had no local zoning ordinances requiring oil and gas facilities to be fenced or marked by warning signs. Following the FCEMD report, county supervisors required new measures to improve security at these facilities (see Section 8.1.7).

---

[23] Forrest County Oil and Gas Well Report. July 2010. The Emergency Management District.

8.1.2   State of Mississippi Regulations

Since 1999, there has been an increase in the number of oil and gas production wells in the state of Mississippi (Figure 8-1).[24] In 2009 there were 5,417 producing oil and gas wells, which were all regulated by the Mississippi Oil and Gas Board (MSO&GB).[25] This is the largest number of wells present since 1951. Multiple wells may feed into a single tank site. The MSO&GB does not collect information on the number of tank batteries in the state. Since 1932, the MSO&GB has regulated the drilling, completion, recompletion and/or operation of oil and gas wells and related facilities throughout Mississippi. The MSO&GB has "sole and exclusive" authority to regulate oil and gas conservation in the state and has "jurisdiction and authority over all persons and property necessary" to enforce all laws requiring the conservation of oil and gas.[26] The MSO&GB has seven inspectors who are responsible for inspecting over 5,000 wells throughout the state.

At the time of the incident, oil field rules promulgated by the MSO&GB did not require fencing, signage, or locks to prevent unauthorized entry to oil and gas sites, apart from those that contain hydrogen sulfide. The MSO&GB does not require inherently safer tank design features be utilized to prevent internal vapor explosions. Rule 6 of the MSO&GB Statewide Rules and Regulations required producers to post a site identification sign with company emergency contact information at tank sites, but did not state where the sign should be located. Additional rules require wells to be located 330 feet from every exterior boundary of the drilling unit;[27] [28] no requirements exist for spacing tanks from existing structures such as residences or public places.

---

[24] MSO&GB Annual Production Report. < http://www.ogb.state ms.us/annprod.php>

[25] The Mississippi Oil and Gas Board. Annual Production Report. Retrieved from <http://www.ogb.state.ms.us/annprod.php. >

[26] The Mississippi Oil and Gas Board. MSO&GB. 53-1-17. Powers of Board. (g)

[27] Drilling unit is the maximum area in a pool which may be assigned to one well to produce recoverable oil or gas.

8.1.3   Communication with the Mississippi Oil and Gas Board

The CSB found the MSO&GB had numerous interactions and communications with personnel from Delphi Oil during the permitting process and the commissioning of drilling operations at the gas well site. There were at least 14 separate communications with Delphi Oil and two site inspections. These inspections did not consider security measures since they were not required in state regulations at the time.

8.1.4   Tanks with Hydrogen Sulfide

The MSO&GB has a site security rule for oil tanks containing hydrogen sulfide in excess of 20 parts per million. Since July 1, 1971, MSO&GB Rule 62 on "Storage Tanks and Sour Crude Oil" has required that oil and gas wells be adequately marked to alert the public and well site workers if a well contains hydrogen sulfide.[29] Additionally, Section A requires that "all access hatches to the tanks capable of being readily operable shall be kept closed securely at all times except when necessary for such hatches to be open for inspection and gauging." The same rule also requires that "all fumes and vapor in such tank or tanks be suitably recovered in a vapor recovery unit or flared to the atmosphere [and] all storage tanks and the nearby surrounding areas be conspicuously marked and posted in a manner advising of the presence of potentially lethal fumes and vapors."

The Safety Practices section of the Mississippi Oil and Gas Statutes, "Operations Involving Hydrogen Sulfide," requires that "safety precaution signs be displayed and unauthorized personnel kept out of the storage area." The CSB determined that similar requirements, if extended to all aboveground production sites storing flammables (with or without hydrogen sulfide hazards), would significantly discourage public access to the sites, preventing possible fires and explosions.

---

[28] MSO&GB. Statutes, Rules of Procedure. Statewide Rules and Regulations. April 3, 2009. Rule 62: Storage Tanks, Sour Crude Oil Pg 88.

[29] *ibid.*

8.1.5   Role of Local Municipalities

At the time of the October 2009 explosion, Forrest County had no zoning ordinances that required fencing, locked hatches, or posting of hazard warning signs at oil and gas sites. However, the City of Laurel, in an adjacent county, had implemented stricter ordinances since 1988 that require fencing and signage at oil and gas sites within the city limits for public safety.[30] In Laurel, fences must be at least six feet high with double stranded barbed wire enclosing all tanks and related equipment.

8.1.6   Forrest County Local Ordinance

As a result of the October 2009 incident, in 2010 the Forrest County Board of Supervisors adopted a local ordinance that requires fencing and warning signs at oil and gas sites that are outside corporate city limits. The measure requires continuous perimeter fencing at least five feet high with one or more strands of barbed wire, locks on gates, and warning and identification signs within five feet of any access point.[31] The ordinance also requires that well operators employ a locking mechanism to "restrict unauthorized use of an exterior gate, door, hatch, ladder, stairway, stairwell or similar device controlling access."

8.1.7   New Tank Storage Site Security Measures: Mississippi Oil and Gas Board

In the aftermath of October 2009 incident in Carnes, in January 2011 the MSO&GB amended Rule 6 of the Mississippi code to require the following at production sites:

- A hazard sign posted at the entrance of well locations in "reasonably large and clear lettering" stating, "Danger," "No Trespassing," and "Authorized Personnel Only."

---

[30] City of Laurel, Mississippi. Code of Ordinance, Ch. 16: Section 75.

[31] Forrest County Board of Supervisors. Ordinance of the Forrest County Board of Supervisors Requiring Certain Safety Measures at Oil and Gas Facilities. July 13, 2010.

- A gate affixed to stairways leading to storage tanks accompanied by a sign reading "Danger (white lettering on red background)," "No Trespassing," and "Authorized Personnel Only."

- The placement of identification signs at site entrances during drilling operations, at the wellbore after well completion, and on the tank battery if it is remote from the wellbore.

- An around-the-clock telephone number posted for reporting incidents at unmanned facilities.

- A sign identifying all wells providing oil or natural gas to a tank battery.

8.2   State of Oklahoma Oil and Gas Rules

8.2.1   Oklahoma Lacks Oil and Gas Sites Security Requirements

The Oklahoma Corporation Commission (OCC) regulates oil and gas site safety for the state. OCC oil field rules do not require a perimeter berm or fencing of oil and gas sites, hazard signs, or hatch locks on oil storage tanks not containing hydrogen sulfide.[32] The OCC does not have any requirements for using inherently safer tank design features. Of the 26 similar incidents identified by the CSB, seven occurred in Oklahoma. Oklahoma reported approximately 41,000 gas wells and 83,000 oil wells in 2009.

8.2.2   Tanks with Hydrogen Sulfide

Similar to Mississippi, the State of Oklahoma requires more public protection measures if oil tanks contain vapor concentrations of hydrogen sulfide. When these conditions are present, OCC rules require warning signs and wind indicators on atmospheric storage tanks. The rules specify language warning about the presence of poisonous hydrogen sulfide.   Signage is required within 50 feet of the facility, to be readable from the entrance and of "sufficient size." The OCC also requires fencing when storage tanks containing hydrogen sulfide above 500 ppm are located inside populated limits of a township or city "where conditions cause the storage tanks to be exposed to the public."

---

[32] Oklahoma Corporation Commission. Chapter 10: Oil and Gas Conservation, July 11, 2009, p. 49.

8.2.3   Tank Hatch Security

The State of Oklahoma changed its security requirements after a fatal accident in 1993 where a 12-year-old boy was asphyxiated upon putting his head into a large hatch of an oil storage tank. On January 1, 1995, Oklahoma state law was amended to require crude oil producers to adopt one of the following security measures:

- Install and maintain a sealing device on the hatch of a tank,

- Reduce the opening of the hatch to less than six inches in diameter or affix, or

- Maintain a sign on or near the hatch no smaller than 40 square inches that warns against opening the hatch and the danger within the storage tank.

The CSB determined that the tanks involved in the April 14, 2010, explosion were equipped with six-inch diameter hatches, but they were not locked.

The OCC also has stronger requirements for certain aboveground flammable storage tanks;[33] however, the rules do not cover the approximately 120,000 oil and gas wells and their associated tanks involved in upstream exploration and production (E&P) activities.[34] Tanks covered under the aboveground storage tank provision are required to be enclosed within a 6-foot high chain link fence, be separated from the fence by at least 10 feet, and have a gate to secure against unauthorized entry.[35] This provision also requires "conspicuously posted" signs with the words "Warning" and "No Smoking" and grounding instructions. CSB investigators determined that the incident in Weleetka would have been less likely to occur if the site were required to follow the fencing and/or warning provisions for either aboveground storage tanks or production sites with hydrogen sulfide hazards.

---

[33] Aboveground storage tanks under OCC Ch 26-1-21 include tanks used in wholesale or bulk distribution activities.

[34] Oklahoma Corporation Commission (2009). Energy, Transportation and Utilities. Annual Report Snapshot FY 2009.

[35] Oklahoma Corporation Commission. Chapter 26: Aboveground Storage Tanks, July 1, 2009. p 27.

8.3   Texas Oil and Gas Rules

8.3.1   Texas lacks oil and gas security requirements

In September 2011, there were over 261,400 producing oil and gas wells in Texas, which were regulated by the Railroad Commission of Texas (RRC). Title 16, Chapter 3, Rule § 3.3, *Identification of Properties, Wells, and Tanks*, requires all oil and gas production facilities to post identification signs displaying the name of the property (as shown on RRC records), the name of the operator, and related information, but does not require fencing, warning signs, or locked tank hatches for tanks without hydrogen sulfide.[36] The RRC does not have requirements for using inherently safer tank design to prevent an internal vapor explosion.

8.3.2   Hydrogen Sulfide Storage Tanks

As in Mississippi and Oklahoma, the RRC has stronger requirements for oil and gas production and storage sites where hydrogen sulfide is present. Under Texas Administrative Code Title 16, Chapter 3, Rule § 3.36, oil and gas production wells with hydrogen sulfide are required to post a warning sign 50 feet from the facility to warn of the dangers.[37] Fencing is also required as a security measure if the oil and gas well site is located inside the limits of a township or city. If the concentration of hydrogen sulfide gas exceeds 100 ppm and the radius of exposure exceeds 50 feet, warning signs must be posted on access roads or public streets. If the concentration of hydrogen sulfide is less than 100 ppm, the lease is not subject to the RRC's hydrogen sulfide rule and lease access and warnings to the public are determined by the operator.

---

[36] Railroad Commission of Texas; Ch. 3, Oil and Gas Division; Rule 3.36, Gas or Geothermal Resources Operations in Hydrogen Sulfide Areas.

[37] Texas Administrative Code Title 16, Economic Regulation; Part 1, Railroad Commission of Texas; Ch. 3, Oil and Gas Division; Rule 3.36, Gas or Geothermal Resources Operations in Hydrogen Sulfide Areas.

8.3.3   State/Municipal Storage Tank Site Security Policies

The CSB reviewed state and local regulations in states with active oil and gas extraction operations, focusing on areas with a high production volume and areas where oil tank explosions have occurred affecting members of the public.

The CSB found a lack of consistent state or municipal regulations for perimeter fencing, hatch locks, and warning signage. The 26 incidents identified by the CSB occurred in ten states. The CSB reviewed the regulations in these states and determined that there is a wide disparity in requirements from state to state. Ohio and California appeared to have the most extensive regulations related to tank security, while some states had no requirements at all. Table 1 summarizes the findings of the analysis.

Table 1: Summary of Oil and Gas Rules in Urban and Rural Areas

Summary of Oil and Gas Rules in Urban (U) and Rural (R) Areas										
Jurisdiction	Fences		Hatch Locks		Warning Signs		Gates		Flame Arrestors		Comments
	U	R	U	R	U	R	U	R	U	R	
California	Y	Y	N*	N*	N	N	N	N	N	N	*Bolted hatches (air emission requirement)
Colorado	Y	N	N*	N	Y**	Y**	Y	Y	N	N	*Gauge hatches to be closed **Prohibit smoking near flammables
Kentucky	N	N	N	N	N	N	N	N	N	N	
Louisiana	N*	N	N	N	N	N	N	N	N	N	*Requires dike/ firewall around tanks in urban areas
Mississippi	N	N	N	N	N	N	N	N	N	N	H_2S sites require fencing, signs and secured hatches
New Mexico	Y	N*	N	N	N	N	N	N*	N	N	*Fencing and gates required for low grade tanks/pits
Ohio	Y	N	Y	N	Y	N	Y	N	Y	N	
Oklahoma	N	N	N	N	N	N	N	N	N	N	
Texas	N	N	N	N	Y	Y	N	N	N	N	Warning signs specific to tank batteries
Los Angeles	Y	Y	N/A*	N/A*	N	N	Y	Y	Y	Y	*Req vapor recovery system and bolted hatches by state
Forrest County, Mississippi	N/A	Y	N	N	N/A	Y*	N/A	Y*	N	N	*Operators restrict use with locking mechanism
City of Laurel, Mississippi	Y	N/A	N	N/A	Y	N/A	Y	N/A	N	N/A	

8.4   Other State Oil and Gas Rules

8.4.1   Ohio Oil and Gas Rules

The Ohio Department of Mineral Resources Management has stronger requirements for tank storage facilities within city limits than in rural areas. Prior to 2004, oil and gas drilling was not permitted in urban areas. However, State House Bill 278 (The Urban Drilling Law) allowed drilling in areas with a population of 5,000 or more people and developed rules to adequately protect the public from the hazards. In urban areas, the law requires oil and gas producers to erect eight-foot-high chain-link fences with three strands of barbed wire around storage tanks, separators, and associated production equipment.[38] The rules require tanks to have spark or lightning arrestors and hatches that are secured at all times. Sites are required to have signs warning against entry and prohibiting smoking. In rural areas, lesser requirements apply.

Figure 8-2: Tank battery in Los Angeles, California

8.4.2   California Oil and Gas Rules

There are over 53,800 producing oil and gas wells in California regulated by the California Department of Conservation, Division of Oil, Gas and Geothermal Resources.[39] However, none of the 26 incidents identified by the CSB occurred in California. The CSB found California was the only state to require some type of fencing in both urban and rural areas for E&P facilities. Title 14 of the California

---

[38] Ohio Department of Mineral Management. 1501:9-9-05 Producing operations.

[39] California Department of Conservation, Division of Oil, Gas and Geothermal Resources. 2009 Annual Report of the State Oil & Gas Supervisor. < ftp://ftp.consrv.ca.gov/pub/oil/annual_reports/2009/PR06_Annual_2009.pdf.>

Code of Regulations requires that all oil and gas equipment located in urban areas be fenced with chain-link perimeter fencing extending a minimum of five feet high with three strands of barbed wire. In rural areas, oil and gas producers can choose whether to install a five-foot chain-link fence as required in the urban areas or a fence constructed of barbed wire or commercial livestock wire netting extending at least four feet high.[40] The rules require identification signs but not warning signs. As a result of more stringent air pollution rules in California, oil and gas storage tanks in air pollution non-attainment areas do not have an atmospheric vent pipe and the hatch is bolted to the top of the tank. Under California clean air requirements enforced by local air resource boards, tank vapors are routed to vapor recovery systems (Figure 8-2). These systems make it virtually impossible to ignite the flammable vapors inside the tank because the atmosphere is too rich to burn.

8.5   Federal Regulations

8.5.1   Federal OSHA Regulations

The federal regulatory framework for E&P facilities includes measures to protect workers onsite and other measures to protect the health and safety of members of the public offsite. The Federal Occupational Safety and Health Administration (OSHA) has regulatory standards that are designed solely to protect workers while the Environmental Protection Agency (EPA) and the Department of Homeland Security (DHS)[41] have regulations intended to protect members of the public outside of facilities. The CSB found no current federal regulatory standards to protect members of the public, including children and young adults, who enter unattended oil sites without authorization.

---

[40] California Code of Regulations, Title 14: Natural Resources, Article 3: 1778. Enclosure Specifications. California Environmental Protection Agency. Air Resources Board. Vapor Recovery Health and Safety Code Statutes. <http://www.arb.ca.gov/regact/2011/evr11/gdfhapp3.pdf>

[41] The Department of Homeland Security promulgated the Chemical Facility Anti-Terrorism Standards to protect U.S. chemical facilities from acts of terrorism. Covered facilities must submit a performance based security plan. DHS does not consider E&P facilities to be a significant risk for acts of terrorism; therefore they do not submit a security plan.

The CSB noted that OSHA has promulgated several standards that include protections for employees working at oil production sites that may also provide a certain degree of protection for members of the public. The most relevant are the storage tank provisions in the Hazard Communication Standard (29 CFR 1910.1200) and the Flammable and Combustible Liquids Standard (1926.152).

8.5.2   The Hazard Communication Standard (29 CFR 1910.1200)

The OSHA Hazard Communication Standard requires employers to ensure that each container of hazardous chemicals in the workplace is labeled, tagged, or marked with the identity of the hazardous chemicals and an appropriate hazard warning.[42] Contrary to the Hazard Communication Standard, many of the oil and gas production sites examined by CSB investigators did not identify the hazardous chemicals contained in storage tanks or provide appropriate hazard warnings.[43] Moreover, OSHA permits employers to use labeling systems with various codes, symbols, and/or numeric ratings that may not be understood by the public. The oil and gas storage sites visited by the CSB did not contain such symbols or numeric ratings.

8.5.3   Flammable and Combustible Liquids Standard (29 CFR 1910.106)

OSHA's Flammable and Combustible Liquids Standard is another occupational regulation that may offer overlapping protections to members of the public. Although the standard is outdated, based on the 1969 version of NFPA 30, it does address many safety issues for aboveground storage tanks including design, construction and installation, corrosion protection, instrumentation, normal vent and emergency relief devices, fire protection, and controlling sources of ignition. However, the OSHA standard has no

---

[42] Occupational Health and Safety Administration. 29 CFR 1910.1200(f)(5)(i)-(ii).

[43] For a detailed discussion and analysis of this issue, see Section 5.3 of the CSB Investigation Report on the Vapor Cloud Deflagration and Fire that occurred at the BLSR Operating Ltd in Rosharon, Texas on January 13, 2003. (Report No. 2003-06-I-TX).

requirements for security or fencing. The standard also exempts crude oil tanks at E&P facilities from requirements for venting valves and flame arrestors.

8.6   Environmental Protection Agency (EPA)

In contrast to OSHA, the EPA has jurisdiction to regulate oil and gas storage sites for the protection of human health and the environment. Accordingly, the agency administers a number of environmental statutes relevant to oil and gas production, including the Clean Air Act (CAA); the Clean Water Act (CWA); the Comprehensive Environmental Response, Compensation, and Liability Act (CERCLA or Superfund); the Resource Conservation and Recovery Act (RCRA); and the Toxic Substances Control Act (TSCA). However, many of these statutes contain various exemptions applicable to oil and gas well sites.

8.6.1   Clean Water Act (CWA)

The Federal Water Pollution Control Act of 1972, as amended, or CWA, is the principal federal statute for protecting navigable waters, adjoining shorelines, and the waters of the contiguous zone from pollution. Section 311 addresses the control of oil and hazardous substance discharges and provides the authority for a program to prevent, prepare for, and respond to such discharges. Specifically, §311(j)(1)(C) mandates regulations establishing procedures, methods, equipment, and other requirements to prevent and contain discharges of oil[44] from facilities and vessels.

8.6.1.1   Spill Prevention Control and Countermeasure (SPCC) Rule

The SPCC regulation promulgated by EPA under the CWA, has been in effect since January 10, 1974 (38 FR 34164). The 1974 SPCC Rule established oil discharge prevention procedures, methods, and

---

[44] Under CWA §311(a)(1), "oil" means "oil of any kind or in any form."

equipment requirements for non-transportation-related facilities with an aboveground oil storage capacity greater than 1,320 gallons (or greater than 660 gallons in a single aboveground tank) or a buried underground oil storage capacity greater than 42,000 gallons. Regulated facilities were also limited to those that, because of their location, could reasonably be expected to discharge oil into the navigable waters of the U.S. or adjoining shorelines. Subparagraph (e)(5)(iii) contains specific requirements for bulk storage tanks at onshore oil production facilities, which are defined in subparagraph (e)(5)(i) as including "all wells, flowlines, separation equipment, storage facilities, gathering lines, and auxiliary non-transportation-related equipment and facilities in a single geographical oil or gas field operated by a single operator."[45]

These requirements address the need for storage tank construction to be compatible with the oil being stored, secondary containment to catch spills, periodic inspection and maintenance, and installation of fail-safe devices to prevent overflow and collapse. Under the security provisions for unattended SPCC facilities, "All plants handling, processing, and storing oil should be fully fenced, and entrance gates should be locked and/or guarded when the plant is not in production or is unattended." However, oil production facilities (see subparagraph (e)(9)(i)), were specifically excluded from compliance with these requirements.[46]

8.6.2   Clean Air Act Amendments

Following a series of chemical accidents in the U.S. and overseas, Congress enacted the Clean Air Act Amendments (CAAA) of 1990. Under 42 U.S.C. § 7412 (r), owners and operators of stationary sources[47]

---

[45] The US Environmental Protection Agency. SPCC Guidance for Regional Inspectors, US EPA, Version 1.1, 3/14/2006 p. 1-2.
[46] 38 FR 34168-70.

[47] Stationary source means any buildings, structures, equipment, installations, or substance-emitting stationary activities that belong to the same industrial group, which are located on one or more contiguous properties and under the control of the same person (or persons under common control), and from which an accidental release may occur (63 FR 645).

must identify hazards, prevent, and minimize the effect of accidental releases whenever extremely

hazardous substances are present at their facility.[48]

This section of the CAAA required the EPA to promulgate an initial list of at least 100 substances that, in

the event of an accidental release, "are known to cause or may reasonably be anticipated to cause death,

injury, or serious adverse effects to human health or the environment." Stationary sources that have more

than a threshold quantity of a regulated substance are subject to accident prevention regulations, including

the requirement to develop a risk management plan (RMP).[49]

E&P facilities were originally considered for coverage under the RMP rule. However, in 1995, the

American Petroleum Institute (API) submitted an analysis[50] to the EPA docket that argued for removing

the facilities from the scope of the rule asserting that the facilities did not pose a significant flammable or

toxic hazard offsite. In January 1998, the EPA agreed to an exemption, stating the "EPA believes

regulated flammable substances in naturally occurring hydrocarbon mixtures,[51] such as crude oil, that

contain many non-volatile components, are unlikely to form large vapor clouds and therefore, generally

have low potential for vapor cloud explosions. EPA considers vapor cloud explosions the consequence of

greatest concern for flammable substances." EPA further stated that "the general duty clause of section

112(r)(1) would apply when site-specific factors make an unlisted chemical extremely hazardous."[52]

---

[48] *Guidance for the Implementation of the General Duty Clause of the Clean Air Act, Section 112(r)(1)*, US EPA, Publication No. EPA 550-B00-002, May 2000, page 2.

[49] 63 FR 640, 640 (January 6, 1998).

[50] Hazard Assessment of Exploration and Production Facilities Potentially Subject to the Environmental Protection Agency's Risk Management Program regulations (API, January 20, 1995).

[51] EPA defines naturally occurring hydrocarbon mixtures as any or all combination of the following: natural gas condensate, crude oil, field gas, and produced water.

[52] 63 FR 642.

8.6.2.1   The General Duty Clause

The Clean Air Act Amendments include a "general duty clause" which holds owners and operators responsible for preventing chemical accidents involving extremely hazardous substances. The clause states that:

> *It shall be the objective of the regulations and programs authorized under this subsection to prevent the accidental release and to minimize the consequences of any such release of any substance listed pursuant to paragraph (3) or any other extremely hazardous substance. The owners and operators of stationary sources producing, processing, handling or storing such substances have a general duty, in the same manner and to the same extent as section 654, title 29 of the United States Code, to identify hazards which may result from such releases using appropriate hazard assessment techniques, to design and maintain a safe facility taking such steps as are necessary to prevent releases, and to minimize the consequences of accidental releases which do occur.*

As part of their responsibility, industries have developed standards and generally accepted safe practices to address the risks posed by extremely hazardous substances.[53] The EPA recommends that owners and operators handling extremely hazardous substances "adhere to a recognized industry standard and practices (as well as government regulations)" to comply with the general duty clause. The EPA advises that when site specific conditions create "unique circumstances" that render some standards "inapplicable," the Agency "may exercise its authority to require an owner or operator to implement additional measures to address the hazard."

To advise the regulated community of its general duty clause obligations, the EPA has published a number of Chemical Safety Alerts. Alerts have addressed a variety of subjects including pressure vessel hazards, lightning hazards to facilities handling flammables, and the catastrophic failure of storage tanks. Security guidance similar to the "Chemical Accident Prevention: Site Security" and "Anhydrous

---

[53] Although there is no definition for extremely hazardous substances, the legislative history of the 1990 Clean Air Act Amendments suggests criteria which EPA may use to determine if a substance is extremely hazardous. The Senate Report stated the intent that the term "extremely hazardous substance" would include any agent "which may or may not be listed or otherwise identified by any Government agency which may as the result of short-term exposures associated with releases to the air cause death, injury or property damage due to its toxicity, reactivity, flammability, volatility, or corrosivity" (Senate Committee on Environment and Public Works, Clean Air Act Amendments of 1989, Senate Report No. 228, 101st Congress, 1st Session 211 (1989).

Ammonia Theft" safety alerts published under the general duty clause would alert operators to the security precautions necessary to prohibit members of the public from entering oil and gas storage facilities. Based on flammability, hydrocarbons stored at E&P facilities would meet the definition of extremely hazardous substances and thus be subject to the CAAA general duty clause.

8.7   Industry Standards and Guidance

The CSB determined that there are currently no comprehensive, specific industry standards or guidance addressing the safety of members of the public at oil and gas sites. The CSB noted some provisions in existing API and NFPA guidance documents that provide limited protections.

8.7.1   American Petroleum Institute (API)

API is a national trade association that represents the oil and natural gas industry and also develops industry standards, recommended practices, and codes.[54] Although they are voluntary, API standards are widely utilized by the energy industry. API standards that are relevant to E&P storage tanks include API 2610 and API 74.

8.7.2   API Standard 2610

API Standard 2610, Design, Construction, Operation, Maintenance and Inspection of Terminal & Tank Facilities (2nd ed.), issued in May 2005, applies to downstream facilities that store refined petroleum products. Section 13.3.6 of API 2610 discusses security measures such as fencing, perimeter lighting, and preventing tank access.[55] The standard recommends that covered facilities be fenced to "maintain facility security and prevent product loss and vandalism," and that "barriers can be added to tank external ladders or stairways to restrict access." The EPA lists API 2610 as a standard that can assist owners and

---

[54] The American Petroleum Institute. About API. <www.api.org/aboutapi>.
[55] The American Petroleum Institute. API 2610, Second Edition, May 2005.

46

operators of SPCC-covered facilities with security plans. However, the scope of API 2610 specifically excludes "[t]anks that are part of oil and gas production, natural gas processing plants, or offshore operations." Nonetheless, the incidents involving public fatalities and injuries at E&P sites demonstrate that these facilities are subject to similar fire and explosion hazards as storage sites in the downstream or refining sector.

8.7.3   API Recommended Practice 74

API Recommended Practice 74, Occupational Safety for Onshore Oil and Gas Production Operation, was developed in response to a CSB safety recommendation that resulted from a 1998 explosion that killed four workers at a Louisiana oil and gas production facility. It includes safety guidance for fire prevention and protection, such as designating areas where there are fire hazards, prohibiting smoking and ignition sources within those areas, posting conspicuous warning signs, and properly labeling tanks that contain flammable liquids. Appendix A of API 74 includes a checklist of questions for periodically assessing safety at oil production facilities.[56] Some questions, for example, suggest that operators verify the posting of "No Smoking," "No Trespassing," and/or "Authorized Personnel Only" signs at oil site entrances. Other checklist questions ask whether "ladders are caged when over 20 feet," if the "access opening to the ladders [is] provided with a swinging gate or chain closure," and whether "tank thief hatches seal or are in good repair."

Beyond these appendix questions, however, the main sections of API 74 include no guidance on requirements for fencing, physical barriers, or security gates to prevent access to tank catwalks and tank hatches; hatch locking mechanisms; or specific tank explosion warning signs to prevent fatal incidents due to unauthorized entry. As currently written, API 74 primarily focuses on occupational safety requirements, containing only limited recommendations for public protection.

---

[56] The American Petroleum Institute. API RP 74, Appendix A, p. 17.

8.7.4   API Oil and Natural Gas Industry Security Assessment and Guidance

Following the 9/11 attacks, the API assessed the E&P sector for security vulnerabilities. The API assessment found most E&P facilities produce low quantities of product; over 75 percent of U.S. oil wells are "stripper" wells that produce fewer than 10 barrels of oil daily. Most are located in rural areas. To provide safety and security, the API suggested the use of the following standards:

- Recommended Practice 49, Drilling and Well Servicing Operations Involving Hydrogen Sulfide

- Recommended Practice 54, Occupational Safety for Oil and Gas Well Drilling and Servicing Operations

- Recommended Practice 74, Occupational Safety for Onshore Oil and Gas Production Operations

- Publication 761, Model Risk Management Plan for E&P Facilities

Of these standards, only API 54 recommends labeling of tanks "to denote their flammable contents" (API 54 Section 8.4.4).

8.7.5   National Fire Protection Association (NFPA)

The NFPA is a non-profit organization that develops and advocates consensus codes and standards for fire protection and prevention. The codes and standards are voluntary unless adopted by law or regulation. The codes are used as good-practice guidance by industry, insurance companies, engineers, and safety professionals. There are three NFPA codes that address security measures for various sectors, but these are not specific to E&P facilities. These codes include NFPA 30, Flammable and Combustible Liquids Code (2008); NFPA 30A, Code for Motor Fuel Dispensing Facilities and Repair Garages (2008); and NFPA 730, Guide for Premises Security (2008).[57]

---

[57] The National Fire Protection Administration. Codes and Standards. Retrieved from <www.nfpa.org/assets/files/PDF/CodesStandards/Directory/NFPADirectory2010.pdf>

8.7.6   NFPA 30

NFPA 30, the Flammable and Combustible Liquids Code (2008), applies to the storage, handling, and use

of flammable and combustible liquids, including waste liquids. Section 21 requires the use of flame

arrestors when storing certain flammable liquids (Class 1B and IC) and Annex A suggests their use to

stop the propagation of a flame inside a tank. Section 22 has requirements for storing liquids in

aboveground tanks including location and installation, normal and emergency venting, fire protection,

spill control, collision protection, and maintenance. In the 1990 edition, the NFPA added subsection 2-

9.3 which states that "unsupervised, isolated aboveground storage tanks shall be secured and marked in

such a manner as to identify the fire hazards of the tank and its contents to the general public." The 2008

edition further states "where necessary, to protect the tank from tampering or trespassing, the area where

the tank is located shall be secured." The NFPA justified the provision based on "several recent tank

explosions caused by youngsters who have trespassed in and on tanks." However, the code has no

specific requirements for security or fencing. The CSB learned that 44 states[58] have adopted a version of

NFPA 30 (ranging from the 1990 to 2008 editions).

8.7.7   NFPA 30A

NFPA 30A, the Code for Motor Fuel Dispensing Facilities and Repair Garages (2008), provides guidance

to reduce hazards of motor fuels from marine/motor fuel-dispensing facilities located inside buildings, at

fleet vehicle motor fuel facilities, farms, isolated construction sites, and motor vehicle repair garages.

Section 4.3.7 contains requirements to physically protect aboveground tanks. If tanks are not enclosed in a

vault, or if the property lacks a perimeter security fence, the code requires a secured gate and a chain link

fence, at least 1.8 meters (6 feet) high and separated from the tanks by at least 3 meters (10 feet). Section

13.3 includes requirements for marking tanks and containers: they must be "conspicuously marked" with

the name of the product and "FLAMMABLE – KEEP FIRE AND FLAME AWAY." The EPA lists this

---

[58] The National Fire Protection Association. Editions Currently Adopted. <NFPA.org>.

standard as a guideline for security for SPCC-covered facilities. However, NFPA 30A does not apply to E&P facilities.

8.7.8   NFPA 730

NFPA 730, Guide for Premises Security (2008), describes construction, protection, occupancy features, and practices intended to reduce security vulnerabilities to life and property. As a guide, this NFPA document is advisory and informational and contains only non-mandatory provisions; the NFPA does not deem the document, as a whole, to be suitable for adoption into law. Chapter 6 discusses requirements for exterior security devices and systems for perimeter protection of facilities and lists detailed specifications for chain link fencing, including design, location of the fence line, signs, height, posts, bracing, top guards, entrances and locks, lighting, and maintenance. Although Chapters 11 through 22 contain specific requirements for different types of facilities (e.g., restaurants, shopping centers, industrial facilities, etc.), there are no specific requirements for oil production sites. However, assuming that such sites can be considered industrial facilities, the guide recommends that access to critical assets be restricted by establishing a secure perimeter accessible only to employees, authorized vendors and contractors, and escorted visitors. Moreover, the guide recommends the following:

> *All industrial companies, big and small, should have site security programs in place to minimize security vulnerabilities and to protect company assets. This is especially true for facilities that handle extremely hazardous substances.*

9.0   CSB Findings

The CSB found that the three incidents in Mississippi, Oklahoma, and Texas, could likely have been prevented by restricting access to the oil and gas production facilities and providing appropriate warning signage, hatch locks, or other appropriate security alternatives; or utilizing inherently safer tank design alternatives. The ease of accessibility and a lack of awareness of the hazards associated with the storage tanks, coupled with the number of oil and gas facilities nationally, demonstrate a potential for similar incidents to occur. The CSB makes the following key findings:

1. Members of the public, most often children and young adults, commonly visit oil and gas production sites without authorization for recreational purposes.

2. Members of the public gain access to production tanks via attached unsecured ladders and catwalks, and may come into contact with flammable vapors from tank vents or unsecured tank hatches.

3. Members of the public, unaware of the explosion and fire hazards associated with the tanks, unintentionally introduce ignition sources for the flammable vapor, leading to explosions.

4. The CSB identified 26 similar incidents between 1983 and 2010, which resulted in a total of 44 fatalities and 25 injuries. All the victims were 25 years of age or less.

5. The three incidents investigated by the CSB in 2009-2010 occurred in isolated, rural wooded areas at production sites that were unfenced, did not have clear or legible warning signs, as required under OSHA's Hazard Communication Standard, and did not have hatch locks to prevent access to the flammable hydrocarbons inside the tanks.

6. The storage tanks did not include inherently safer design features to prevent tank explosions. Safer design features used in the downstream, refining sector would likely prevent tank explosions at E&P sites. These include the use of vents fitted with pressure-vacuum devices, flame arrestors, vapor recovery systems, floating roofs or an equivalent alternative.

7. E&P storage tanks are exempt from the security requirements of the Clean Water Act and from the risk management requirements of the Clean Air Act.

8. Industry guidance from the American Petroleum Institute recommends specific security measures for storage tanks of refined petroleum products but not for storage tanks at upstream E&P sites, and the National Fire Protection Association standards do not adequately define security expectations where these deadly incidents occurred.

9. Some states, including California and Ohio, and some localities have mandated security (including fencing, locked or sealed tank hatches, and warning signs) for E&P sites, particularly in urban areas. As a result, despite its large role as an oil producing state with many of these types of facilities, none of the 26 incidents occurred in California. However, many other large oil and gas producing states have no such requirements (except for certain E&P sites where toxic hydrogen sulfide gas is present).

10.0 Recommendations

The CSB makes the following recommendations:

The Environmental Protection Agency

2011-H-1-R01

Publish a safety alert directed to owners and operators of exploration and production facilities with flammable storage tanks, advising them of their general duty clause responsibilities for accident prevention under the Clean Air Act. At a minimum, the safety alert should:

  a) Warn that storage tanks at unmanned facilities may be subject to tampering or introduction of ignition sources by members of the public, which could result in a tank explosion or other accidental release to the environment

  b) Recommend the use of inherently safer storage tank design features to reduce the likelihood of explosions, including restrictions on the use of open vents for flammable hydrocarbons, flame arrestors, pressure vacuum vent valves, floating roofs, vapor recovery systems or an equivalent alternative.

  c) Describe sufficient security measures to prevent non-employee access to flammable storage tanks, including such measures as a full fence surrounding the tank with locked gate, hatch locks on tank manways, and barriers securely attached to tank external ladders or stairways

  d) Recommend that hazard signs or placards be displayed on or near tanks to identify the fire and explosion hazards using words and symbols recognizable by the general public

The Mississippi Oil and Gas Board

2011-H-1-R02

Amend state oil and gas regulations to require the use of inherently safer tank design features such as flame arrestors, pressure vacuum vents, floating roofs, vapor recovery systems or an equivalent alternative, to prevent the ignition of a flammable atmosphere inside the tank.

Oklahoma Corporation Commission

2011-H-1-R03

Amend state oil and gas regulations to:

  a) Protect storage tanks at exploration and production sites from public access by requiring sufficient security measures, such as full fencing with a locked gate, hatch locks on tank manways, and barriers securely attached to tank external ladders and stairways.

b) Require hazards signs or placards on or near tanks that identify the fire and explosion hazards using words and symbols recognizable by the general public.

c) Require the use of inherently safer tank design features such as flame arrestors, pressure vacuum vents, floating roofs, vapor recovery systems or an equivalent alternative, to prevent the ignition of a flammable atmosphere inside the tank.

The Texas Railroad Commission (RRC)

2011-H-1-R04

Amend state oil and gas regulations to:

a) Protect storage tanks at exploration and production sites from public access by requiring sufficient security measures, such as full fencing with a locked gate, hatch locks on tank manways, and barriers securely attached to tank external ladders and stairways.

b) Require hazards signs or placards on or near tanks that identify the fire and explosion hazards using words and symbols recognizable to the general public.

c) Require the use of inherently safer tank design features such as flame arrestors, pressure vacuum vents, floating roofs, vapor recovery systems or an equivalent alternative to prevent the ignition of a flammable atmosphere inside the tank.

American Petroleum Institute

2011-H-1-R05

Create a new standard or amend existing standards covering exploration and production facilities to:

a) Warn that storage tanks at unmanned facilities may be subject to tampering or introduction of ignition sources by members of the public, which could result in a tank explosion or other accidental release to the environment.

b) Recommend the use inherently safer storage tank design features to reduce the likelihood of explosions, including restrictions on the use of open vents for flammable hydrocarbons, flame arrestors, pressure vacuum vent valves, floating roofs, vapor recovery systems or an equivalent alternative.

c) Require security measures at least as protective as API 2610 to prevent non-employee access to flammable storage tanks at upstream E&P sites, including such measures as a full fence surrounding the tank(s) with a locked gate, hatch locks on tank manways, and barriers securely attached to tank external ladders or stairways.

d) Require that hazard signs or placards be displayed on or near tanks to identify the fire and explosion hazards using words and symbols recognizable by the general public.

e) Recommend that new or revised mineral leasing agreements include security and signage requirements as described above.

The National Fire Protection Association

2011-H-1-R06

Amend NFPA 30, "Storage of Liquids in Tanks—Requirements for all Storage Tanks" as follows:

a) Remove the term "isolated" from the current wording of the standard and replace it with a more descriptive term, such as "normally unoccupied"

a) Remove the words "Where necessary" from Security for Unsupervised Storage Tanks, Chapter 21.7.2.2.

b) Add a reference to a relevant security standard that offers specifications on fencing, locks and other site security measures.

c) Add a definition of security encompassing requirements such as fencing, locked gates, hatch locks, and barriers.

By the

U.S. Chemical Safety and Hazard Investigation Board

> The Honorable Rafael Moure Eraso
> Chair
>
> The Honorable John S. Bresland
> Member
>
> The Honorable Mark Griffon
> Member

Date of Board Approval

11.0 Appendix A: Previous Incidents

Rio Blanco County, Colorado, June 24, 2007: 2 Teen Fatalities[59]

On June 24, 2007, in Rio Blanco County, Colorado, a tank exploded when a group of 15 to 20 teens and young adults were socializing near an oil tank storage site. The site was located along an access road on public land leased from the National Forest Service. The production site had two 400-barrel capacity tanks with atmospheric vent pipes; one contained 60 barrels and the other contained 120 barrels of oil. The youths had built a campfire approximately 50 to 60 yards from a group of oil tanks and two or three individuals and a dog climbed onto the tanks. Witnesses indicated that these individuals were jumping on the tanks when the witnesses heard a hissing coming from the tanks. Approximately 10 minutes later, an explosion propelled the bottom of one tank 80 to 100 yards away from its original location. The force of the explosion killed the two teens jumping on the tanks. The oil site was located in an isolated area and the embankment surrounding the tanks had a steel construction type pipe fence around it, intended to restrict cattle. The Colorado Bureau of Investigations (CBI) determined a lighter to be the likely ignition source and recommended that fencing be placed around the pumping unit and pit area.

Mercedes, Texas, May 17, 2007: 3 Teen Fatalities[60]

A tank explosion on May 17, 2007, in Mercedes, Texas, killed three teens. Shoe prints were found on top of the tank, indicating that the teens were likely on top of the tank prior to the explosion; a cigarette lighter was also found. The tank was easily accessible and was a regular hangout for the teens.

---

[59] Colorado Bureau of Investigations

[60] Hidalgo County Fire Marshal's Office

Long Lake, Texas, April 11, 2003: 4 Teen Fatalities[61]

A tank explosion on April 11, 2003, in Anderson County, Texas, killed three teens instantly; another died later of his injuries. Five teens had gathered at a remote oil site to socialize. They climbed the catwalk that accessed several tanks. They climbed to the top of the tank when one teenager climbed back down to the catwalk, opened an access hatch to one of the tanks, and looked inside. He returned to his spot on top of one of the tanks while one of his friends climbed down to look inside the tank, using a lighter to see the contents more clearly.   The tank exploded fatally injuring three, seriously injuring one teen and leaving one with minor injuries. The site had no fences or warning signs.

Heflin, Louisiana, May 26, 2001: 1 Teen Fatality [62]

An oil tank incident on May 26, 2001, in Heflin, Louisiana, killed one teen, severely burned another, and left a third with minor burns. Six teens gathered at an oil site at approximately 5:00 a.m., when three of the six climbed the catwalk of a tank, which allowed access to the tops of three oil tanks. One teen climbed the center tank; another joined him, but became frightened while on the tank and climbed back down to the catwalk. The teen on top of the tank was smoking a cigarette, which likely ignited flammable fumes venting from the tank, causing the explosion. The force from the explosion threw the teen approximately 96 feet from the tank, killing him. Another teen on the catwalk was doused in burning liquid and severely burned. The third received minor burns.

---

[61] Anderson County Sheriff Department
[62] Webster Parish Sheriff's Office

	Oil and Gas Storage Site Explosions, 1983-2010						
	Date	City	State	Fatality	Injury	Incident Summary	Fencing Required
1	4/2/1983	Center	TX	2	0	Explosions blew apart two storage tanks of gas distillate killing two fourteen-year-old girls playing nearby; four other youths escaped uninjured.	No
2	6/24/1985	Centralia	IL	1	4	Firecrackers tossed into a 10-ft oil tank triggered an explosion and killed one teenager and injured four, including the father of a victim. The blast sprayed crude oil on homes nearly 100 yards away.	No
3	5/16/1990	Beggs	OK	3	0	An oil storage tank exploded, killing three men in their early 20s when one of the men attempted to light a cigarette.	No
4	8/19/1990	Logansport	LA	4	0	Four people, including two teenage sisters, were killed when a storage tank exploded. The victims appeared to be climbing a ladder to the top of the tank when the explosion happened.	No
5	6/19/1991	Oklahoma City	OK	1	0	A 13-year-old boy playing atop an oil field salt water tank was killed by an explosion triggered when he apparently struck a match. His body was thrown 60 yards by the blast.	Yes (Urban)
6	10/28/1991	Tyler County	TX	0	2	An adult and a teenager were injured when an oil storage tank exploded, igniting four other tanks when one of the victims lit a cigarette lighter near the storage tank's open hatch.	No
7	9/22/1992	Sherman	TX	1	4	A tank explosion killed a teenager and injured four others at a tank farm where seven teenagers were partying after four teens climbed a tank and lit a match to see inside.	Yes (Urban)

	Date	City	State	Fatality	Injury	Incident Summary	Fencing Required
8	7/2/1993	Providence	KY	4	5	Four teenagers were killed and five others injured in a crude oil storage tank explosion triggered by a cigarette at a party. The victims were sitting on top the tank when it exploded.	No
9	4/23/1995	Duncan	OK	3	0	An oil field blast killed three 13-year-old boys while playing near two remote oil field storage tanks.	No
10	11/28/1995	Bradford	PA	2	0	A lit cigarette was blamed for an oil tank explosion that killed two 14-year-old boys playing on the tanks.	No
11	6/22/1997	Konawa	OK	2	0	Two teenagers, 15 and 13, died when they climbed a 20,000-gallon oil tank and lit fireworks, triggering an explosion.	No
12	7/29/1997	Chandlersville	OH	2	0	Two teenagers, 17 and 15, died when a 15-ft oil tank exploded, throwing their bodies more than 200 feet. The teens appeared to have been climbing the tank at the time of the explosion.	No
13	8/14/1998	Logan	OH	1	1	One young man was killed and a 16-year-old girl was seriously injured when an oil storage tank exploded, throwing their bodies about 100 feet. The oil tanks were not in use at the time of the explosion.	No
14	1/29/2000	Flora Vista	NM	1	1	A gas tank explosion killed one teenager and critically injured another when one of the boys apparently threw a lighter into the 12,000-gallon tank.	No
15	5/26/2001	Sibley	LA	1	1	One teenager was killed and another critically burned when an oil tank in an oil field exploded. Two of the five teens were sitting on the tank, and one was smoking a cigarette at the time of the explosion.	No

	Date	City	State	Fatality	Injury	Incident Summary	Fencing Required
16	11/30/2001	Duson	LA	0	1	An explosion in a crude oil storage tank threw a 14-year-old boy more than 100 feet as he was reportedly walking his dog in the surrounding fields.	No
17	4/11/2003	Long Lake (Palestine)	TX	4	0	Four teenagers died in an oil storage tank explosion when five teens climbed the tank and one opened the hatch. A cigarette lighter triggered the blast and the victims were thrown nearly 75 yards.	No
18	9/6/2003	Blue Rock	OH	0	2	Four individuals were socializing at an oil tank site when a tank exploded causing head injuries to two of the men. One of the men apparently lit a cigarette after climbing atop the tank.	No
19	5/14/2005	Ripley	OK	2	0	Two men ages 19 and 20 died from third-degree burns over 90 percent of their bodies after an oil storage tank exploded while they and two others were drinking at the site.	No
20	12/18/2006	Springtown	TX	1	1	Two teenagers, 16 and 14, at a tank battery dropped a burning paper into an unlocked tank hatch located inside a 5-ft high unlocked cattle fence. One victim was killed and the other injured.	No* (cattle fence)
21	3/12/2007	Oklahoma City	OK	0	1	A 15-year-old boy was critically injured after an explosion and fire at an oil tank battery burned more than 45 percent of his body. The cause of the explosion was unclear.	Yes (Urban)
22	5/18/2007	Mercedes	TX	3	0	Three teenagers were killed when a liquid storage tank exploded in a field after one of the teens apparently climbed onto the abandoned tank and opened the hatch.	No

	Date	City	State	Fatality	Injury	Incident Summary	Fencing Required
23	6/23/2007	Oak Creek	CO	2	0	A group of 15-20 teens partying at an oil storage site triggered a tank explosion killing two teens as they jumped on and smoked near the oil tanks.	Yes (Wildlife)
24	10/31/2009	Carnes	MS	2	0	Two teenagers socializing at a tank site were killed when an oil tank exploded in a wooded clearing approximately 150 yards from one of the victims' homes.	No
25	4/14/2010	Weleetka	OK	1	1	A group of 6 teenagers and young adults were socializing at an oil storage site when a tank exploded, fatally injuring one and causing second degree burns to another.	No
26	4/26/2010	New London	TX	1	1	Two young adults were socializing at an oil tank site when an explosion killed one and critically injured the other.	No

Sources: 1) Associated Press; 2) Chicago Tribune; 3) USA Today; 4) Washington Post; 5) Daily Oklahoman; 6) San Antonio Daily Express; 7) New York Times; 8) Associated Press/CSB Documents; 9) Daily Oklahoman; 10) Pittsburgh Gazette; 11) Associated Press; 12) Columbus Dispatch; 13) Associated Press/CSB Documents; 14) Albuquerque Tribune; 15) Associated Press/CSB Documents; 16) Daily Advertiser; 17) Associated Press; 18) Associated Press/CSB Documents; 19) Associated Press; 20) CSB Documents; 21) Daily Oklahoman; 22) Associated Press; 23) CSB Documents; 24) CSB Investigation; 25) CSB Investigation; 26) CSB Investigation